Why Photographers Commit Suicide

Why Photographers Commit Suicide

Poems by
Mary McCray

Trementina Books

Books by Mary McCray

St. Lou Haiku

Contents

Introduction by Howard Schwartz   1

Preface   3

Imagine Mars   7

Traveler, Beware   8

The Torus City of Viseratonia   10

Helga Traveling   11

Bluestone   12

Things My Astronaut Is Afraid Of   15

The End, Part I   16

Sex in Zero Gravity   18

Building Mars   19

Genealogy of the Stars   20

Phineas Revisited   24

The Birds of Mars   26

Sex with a Rocket   29

Hardly Rocket Science   30

The End, Part II   31

Helga in the Park   34

Sacajawea, I'm Just a Fan   35

Touring Thorazine   37

Failure to Launch   39

Profiling Davidson, Initial Report   41

Earth Watchers   42

Jane and Maurice   43

All of a Sudden   47

Starbaby   48

On a Clear Day You Can See Jupiter   50

Maribel, Crossing   51

Helga Post-Orbit   52

Children of Algebra   55

Um Is A Wonderful World   56

Monogamous Carbon, a Classified Ad   58

Things My Inner Astronaut Is Afraid Of   61

The End, Part III   65

Y   66

Why Photographers Commit Suicide   67

Acknowledgements   71

for curiosity & melancholy explorers

Introduction

Mary McCray is a trailblazer in science fiction poetry. *Why Photographers Commit Suicide* is an ingenious vision of a future in which life on Mars resembles life on Earth as we know it, although it is also like living in a test tube. Presented in a voice as natural and relaxed as an afternoon at the country club, these poems are also a satire of life on this planet: "Come see Mars…You'll be bargaining for artifacts/ drinking Martian Margaritas/and writing free verse."

There is also some wonderful personification of the solar system: "A couple of stars, she and her husband/revolve around each other/ so deferentially." In this vision of the future so much like the present, every effort has been made by the authorities to transform the barren Martian desert into a familiar landscape.

McCray cleverly compares the new world of Mars with that discovered by Lewis and Clark: "A new world spread/like a blanket over the old." Even living birds have been imported who reliably "start waking up souls." Holograms of popular figures, including Sacajawea who has been resurrected as a pop singer, walk through crowds. One fan notes, "Sacajawea, how empty space is/when you go off the air."

Focusing on "the impossible mysteries of Mars," McCray not only explores life on Mars, but the life of those left behind on a more familiar planet, our own. This unusual collection heralds the arrival of a poet whose vision is distinctly unique and fascinating.

Howard Schwartz, Summer 2012

Preface

Poetry is organized thought, meticulously organized thought. A poet maneuvers and organizes the many modes of thinking: subconscious, logical, metaphorical and metaphysical.

The best quality of poetry is a smooth combination of all forms of thinking and although there is what we call an art and a craft involved, writers are only as good as their best thoughts, their best deductions, their best theories, and their most amazing connections between disparate ideas.

Although I am not a scientist or very strong with numbers, I have always felt there was a scientific process to a poet's thinking and I have always loved to watch human beings working through their strategies and predicaments, be they detectives putting together clues in a movie, chemist-chefs fixing recipes for a dinner party, psychologists trying to make sense out of a zeitgeist, mathematicians crunching numbers for Sudoku, or a designer creating an entire outfit out of old typewriter ribbon. None of these efforts strike me as any more or less interesting than a human mind organizing ideas and solving problems in verse.

Problem solving is our common task. Every discipline of our universe has hurdles to clear. And we all use subconscious, logical, metaphorical and metaphysical thinking to overcome our challenges.

In light of this commonality, I hope one day all disciplines of thinking, all arts and sciences, will become reconciled to each other, that geologists and oceanographers will read poems to research the mysteries of land and sea, and that poets will embark on voyages to locate sunken ships, to decipher the physics of music, to unearth the charms of a treasure map, or to explore space.

Mary McCray, Summer 2012

Imagine Mars

Imagine your life in a glass box,
cold, clean streets and the porcupine feeling
of antiseptic air.
Imagine holy days of rain,
a cyclone, staged burlesque
in a carnival.
Organized animal love.
Imagine everything Mars,
a world without squalling,
the gauzy breeze tribesmen tell of
in Legend of the Faucet Sky.
Imagine the smell of autumn in a test tube,
cloning sickly trees,
sheet-snapped flurries with nowhere to go,
leaflessness.

Traveler, Beware

If you find yourself driving, lost
and the October trees outside
your Chevette are wrestling
with their memories of daylight

and because of astigmatism
in your one good headlight
you can't see ahead,
only hear the ring-chink
of a bell on a bicycle behind you
and the maniacal laughter so suited
to your child, your once-was-in-you
child—

whoever that was,

if you feel your spine suddenly
and you can't look back—

whoever that is,

and you think you may as well be
adrift putting through the outer
atmosphere until you recognize
the side rails swooning off the edge of the road
and the roadless, yellow lifeline
luring you into the motorless night,
impossibly past the abandoned
Volvo of your lover,

dome light on, rear door left
ajar

and the voice of the asphalt road
with its chorale of monotone:
thumping *this way, this way*

and finally, the dangerously roundabout question
comes from the overlit truck stop—

where are you now?

or the blinding headlight
of dawn asking—

what happened?

The Torus City of Visceratonia

Charlie calls me on the phone one day and he says,
"Come see Mars." And his wife Elizabeth says,
"Yes, you simply must tour Visceratonia."
"Meet us by the ancient cave cities,"
Charlie says, "off the cracked rim
of the southern solar shell.
Soak in some acceptable rem
with the colorful East-enders;
They spend romantic lives sweating on tugboats,
pulling asteroids in for cosmic carbon."

"You'll love the culture clubs," Elizabeth says,
"the electric subway
poetry, the out-of-water fish.
At first you'll miss the toilets
in the artificial gravity spin.
But you'll be mesmerized
by the streamy gyrations
of the synchronized sprinklers.
Soon you'll be clocking rainbows
and watching terraforming over tea.

You'll be bargaining for artifacts,
drinking Martian Margaritas
and writing free verse."
"You'll be a traveler in the Martian century,"
Charlie says, "surfing in the solar wind."
"They live frontier lives out here," Elizabeth said,
"wild expatriate lives."

Helga Traveling

Sneaking Marge's dog into space wasn't easy.
Not like that cow they tried to sneak out
on a cargo shuttle back in '62.
As you can imagine, it's a hard thing
to sneak a cow into space,
what with the weight upon the axis of the machine
and the Holstein breath fogging up the portholes.

Marge tried every legitimate way to get Helga off Earth.
She filled out forms, passed tests, negotiated various options
of quarantine, wrote 36 letters to her representative,
a 400-pound man repeatedly denied passage himself.
She even tried to wrangle a working-dog visa.

But in the end, she was stashing mainstay chewies
in her pant pockets and poking holes in a box.
Then on Mars, we found her praying over her suitcase
like a Hare Krishna, counting pink toes and vampire teeth,
waiting for the zero-gravity grin on those black, gristly lips
and the white wagging tip of a plush tail-star.

Bluestone

If you had been the last to leave Earth
you would have had all the Earth stuff:
backyard weeds
battered capes
gently rolling plains
crawling with nothing
you would have had all the marbles
a cat's eye in red mud
cathedral forests
ashes of eroding
footprints
(because a billion will let a treasure go untreasured)
the last   freehold   of   o x y g e n

Things My Astronaut Is Afraid Of

- Talking about our collected net worth
- Coming in outer space
- Raindrops on roses
- And reckless zambonies
- Crowds that surround a clown when he sings
- Redundant neurosis and birds with one wing
- Herds of imitation Elvi
- Embracing my revolutionary belief system
- Entropy, entropy, entropy and
- Sleeping alone

The End, Part I

The melodramatic monologue of a jilted planet

Once every million years you lose a date, one of those days
when love and an icy comet of bad chemistry
blows you off your horse.
The bulbs of shooting stars go bust and narratives fall apart.

Doesn't take much, a curt wave from a window,
nervous cough over a telephone.
You suddenly catch the smell of your life
with less than a second to respond, smoldering.

On the last day of the last month of the last year
I, a very jealous and depressed planet, was caught
measuring my life. All the way to the end,
rejected, bereft of everything but the words
'swept off my feet.'

You see, I was deeply into 363 hermetic seconds of lunar outpost
        woe,
when God looked down on me and said
"I'm not in love with you anymore."

And interestingly,
the laws of complex panic hold universally.
Rules on an unsteady tempest boat do apply
to the endless end: let go and you are lost,
hang on and you are beaten.

If you've ever been dumped right out of your shadow,
cuckolded out of your life, there at the end
when the wreckage of a twisted universe spits you out,

there is nothing more to wrestle over (God and "I don't love you
    anymore").

I muscled over your pictures every night for a week—
your landscapes, love lines, water gaps.
Can I start all over again?

Sex in Zero Gravity

astronaut, astronaut—
kiss me with your incomplete sentences
and your raw relativity,
run your fingers like lasers,
escape velocity through my motor heart,
the acceleration thrust
of your deep-space Cadillac cruising
my jellyfish tremors,
touching the swirling hurricane
that is the red G-spot of Jupiter,
my electron shifting off its shell
in a quantum leap of heat,
a roll of light that comes, warp speed,
into the whole ocean of my body,
channeling disorder with a drop of ink
in water, a calming sense of insignificance,
like touching flakes of snow
that disappear before you know
you've just had your arms around God.

Building Mars

Little boys play with wheelbarrows
during the very first days of settling Mars.
Little girls push Tonka trucks
around strewn lumber and tug backhoes
through the loam to the quarry
where derricks and cranes and Matchbox rammers
run through the rocky red mud.
A sunburned foreman dumps his Lincoln Logs
and a tremor moves the architect, head bent
with crayons and small hands in his plots and schematics.
Engineers put together Tinker Toys with putty.
Union men excavate the mountains
with jackhammers, start stacking Legos.
In the evenings, a tarpaulin settles over adult couplings,
nails, sockets, screws, fixtures, slide-rule adrenaline.
Flood lights swing from scaffolding
and children follow the steeplejack
up through the crossbeams into a sleepy temple
overlooking a mud-red empire.
Out of a dirt-stained window they see
GI Joe foreman stepping out of his trailer.
Around the yard he surveys and bids
on premium space; gets time and a half for raising
an airtight Martian metropolis out of dirt.
His skyscrapers, bridges, and fountains,
his muddy toy city so simple and small,
any mischievous seven-year old ruffian
with his thumb-size wrecking-ball
could knock it all down.

Genealogy of the Stars

I. Alice was never close to her parents,
   their lines of sight being strained
   and their orbits all askew.
   Father Firmament is a thing as old
   as the Milky Way. He sits back
   in his easy chair smoking helium, gray swirls
   whisper around his head like a crown.
   Mother Beehive rocks next to him,
   back and forth, sewing the galactic circle.
   Most nights her folks are just hazy spheres
   out in the outskirts, the boonies
   of this midnight county, so far out
   only the fingers of ham radios
   reach them at Omega Centauri—
   Route 6, Box 305.
   She was born there
   at the rim of the pasture,
   by the tool shed, shack of our Lord,
   too bottomless and creepy
   for even the flashes of Polaroids.

II. Someday she will have seven
    children of her own—
    Rita, Wilomena, Constelleone,
    Berta, Roberta, Shasta and Osh,
    seven little women swinging
    on the swings of gravity, flying
    carousel orbits with tens of thousands
    of their little winking friends.
    Giggling in the bedrooms of Taurus,

they smirk at the old, overheated suns
while plastic glow-in-the-dark stars
hang over their midnight cribs.
In the morning, the Pleiades girls
are flipping marbles into the games
of boys. They sit in highchairs
with their feet hanging over
toy planets and the shag, ebony floor.
Once a year the dog star races through
their cul-de-sacs and the shredded leaves
of asteroids blow through the yard.

III. Once upon a time, Alice married
    into the Gamma Leonis
    and that's what the mailbox says.
    A couple of stars, she and her husband
    revolve around each other,
    so deferentially.
    Every four hundred years,
    once in a hurly-gone while
    you will see them and their corridor waltzes,
    their bickering eons, back and forth,
    their midlife-crises, their fear of falling.
    Just wipe your telescope lens
    and show them your gargantuan eye,
    pupil gawking night after night,
    grass growing in between your toes.
    Note the trembling of their constellation
    heartbreak, their Velcro devotion
    in the face of the felt abyss,
    all their hanging on.

Phineas Revisited

Back in 3014, Stephen Sharp and Jack Wittinger of Local 656 were laying tubing for the tracks of the Ronks & Riverside Line, an airbus from the Mars city Myopia to the Outback Biosphere, when a spontaneous detonation of dynamite shot a plastic rod through Stephen's head. It was just like Phineas Gage back in Earthdate 1848. An alien object tunneled through his brain just under his left cheekbone, right out through the top of his head, obliterating his left eye. Stephen just stood there, unfazed, swinging his head and that rod back and round. It made you chuckle before the horror of it sunk in.

The crew walked him over to the Ox Shuttle and flew him over to First Aid. Stephen kept telling the doctor he didn't want to lose any pay. As it turned out, he got fourteen weeks of Worker's Comp. After ten weeks of therapy, he regained an ability to live, a general walking, talking, breathing routine. Ronks & Riverside hired him back, but the well-liked union man, once an efficient and capable lineman, was gone. Stephen was now sloppy, neurotic, violent and prone to graffiti. As the rest of Local 656 used to say, "He was no longer Sharp." It was as if some slice of his brain-pie that had held the whole of his personality had been removed. A planet of useless hemispheres was left, total destruction of a thing known as *self*.

A team of doctors commissioned a study on Stephen. With their laboratories and flasks, they devised theories with words like "neocortex" and "temporal lobes." In the end, they gave up and Stephen was given a lobotomy.

Stephen's story is chock-full of science: biology, chemistry, arithmetic, electricity. We have no grip on most of it, the drywall of our peculiarness, the substrate of our selves. Who you are. The doctors said that the brain is a machine that filters but does not generate *you*, that there is no one part of Stephen's brain that was his *I*.

Self is just an illusion we need for survival. How precarious we are, then, on the precipice of our personalities. Enough to give you haunting dislocation, a true sense of vertigo. *I* think Stephen's personality is out there sailing through outer space, a refugee floating outside the main ship, small and shipwrecked, shuttling to dock on another planet-person *I* — which would explain multiple personalities then, wouldn't it?

The Birds of Mars

Imported for the lifeless Martian morning, the small beaks
of living birds fog circles in the windows of a watchmaker.
Mars rocks, spurts, and wiggles on its pendulum.
And all over the ground, the birds hop, dependably full of atoms.

Near a street corner, a man stands winding his grandfather's watch;
maybe these birds are in his bloodstream, scratching at his heart;
the cogs and wheels and valves of that punctual chest muscle
never losing a second, dodging the bullets of friends,
working three billion times without fail,
ticking more perfect than time.

And as sand runs down through the core of the planet,
the birds flutter in the air; their claws go to the wires.
They cock their metronome heads
to thin crystal-quartz chirping.

With all the guts of their technology,
feathered timekeepers do not understand clocks.
They do not understand the coiled-spring inaccuracy
of seconds. They match their cogs and wheels
to the slowing, tipping planet,
conform their time to an earthquake pulse,
to the sundial shadows and the gradual
billionth-of-a-second burning out of days.

For them, no clock exists
but the planet's imprecise and awkward revolutions.
And so ever reliably, in an imperceptible break of light,
the sparrows land the morning and start waking up souls.

Sex with a Rocket

O Astronaut, my fellow traveler,
I am letting go of you here at the launch.
You must fall back like a rocket pack,
your hot breath floating up
in a bubble into the ocean of night.
The blast-off feels tree-like, an elm torpedo,
water rushing through my leaves,
a radioactive trunk swirling
into a slow swirling rub
swelling bark bumps.

Is someone else in the water,
swimming in the shock of sudden nothingness
with the translucent arms of an octopus
holding me poised with my belly up high?

It is here all our ambitious missions dissipate
like salt seeping into my skin
like the osmosis of an emotion.

Hardly Rocket Science

The story of meteorite ALH84001 who fell to Earth

I sit on ice like a small potato.
I am the Mars rock exiled in some astro-boy's Frigidaire.
But it hardly counts—being the banished stone
from some hard-hearted globe.
I have rolled through space and sixteen million years
to the ice fields of another planet,
to the exploding breakdown of my life.
It's hardly calculus, my literal alienation.
Bagged in 1984, I cooled my heels until 1993
when some politician's hooker called me
a vegetable of Pluto. Now I'm famous.
But it's hardly worth the fuel, the energy spent
breaking through the atmosphere.
I was just a sentimental little rocket
falling apart in slow motion,
in front of the bleachers and all of my friends,
shaved pieces of me plunging into a sea of spectators,
forty-two penguins waddling through Alan Hills, Antarctica.
My mother cried.
On TV, the boys of science said
it was all remainders gone wrong:
binaries adding up to leftovers,
leftover rocks. leftover parts,
misfits shifting to the ends of kickball teams,
scraps at parties, friends by default,
the curses of long division.
Astro-geologists argue over me.
Some call me proof of life and Martian biology.
Others call me paper weight.

The End, Part II

At last, ladies and gentlemen, physics.
Darkness reports the Sun, alias the Star,
long believed to be "under the weather"
due to a running out of fuel necessary
for nuclear fusion in its core,
has finally succumbed to its own gravity,
(the weakest of all four elementary forces),
and has collapsed.
We have no comment from the supernova,
but a *black hole* is expected to be imminent.

Meanwhile, miscellaneous rock and dust have been speeding
forty-thousand miles per hour
through the new star-barren Milky Way abyss.

　　　CUT TO: hurling cloud of dust.

God was spotted, possibly erroneously, as a roach
toiling on one of the two Martian moons
and is said to have said the following:
　　　TEXT OVERLAY: "Finish your arithmetic assignment."
We are not yet sure what prophecy this new passage contains
but our staff is currently working extended hours to crack the code.
The immediate capture and autopsy of the roach reveals
definite traces of Alpha Centauri in its bloodstream.

A spokesperson from Andromeda, the next-door galaxy,
had this to say: "Stars die. They burn out everyday.
You can't let yourself get emotional about it."

Witnesses described the star's last few moments as "hectic,"
"like a spatial hurricane" and "hot."
One obviously shaken informant said he could actually
"feel the universe stretching."

In the following footage taken from the scene,
you can see our ancient Sun this morning
as seen right before she blew out.
      CUT TO: the sun in a zen state of contentment.
And here we have a picture of a star that died months ago
and has now cooled as a black dwarf.
Our Sun, however, is expected to become a *black hole.*
I repeat, the latest information seems to confirm
that we are to be sucked into a *black hole.*

As a reminder, no bang, no light, no radiation,
no matter can escape the infinitely dense apartment
of a *black hole.*

Is there a universal law?
Do you ever wonder about the meaning of life?
We'll explore these issues and more in our ten o'clock edition.

And later tonight, as the end of light folds into the pit of
    nothingness,
we will air for the last time in time,
a movie based on the true-life events of this story.
Here's a clip: *EXTERIOR: darkness,*
        a primordial ocean of gas and stars.
        Enter disembodied voice-overs, schizophrenia,
        consciousness coming apart at the seams.
        CUT TO: ten nanoseconds of therapy
        failing to cure these strange impulses.

A curtain of helium ashes falls
upon a desert of quiet, dark matter.
PAN ACROSS IMMEASURABLE DAMAGE.

Tune in then. That's all for now.
The world is watching…Today more than ever.

FADE TO STILL WATER.

Helga in the Park

Mean as bad traveling and her ears are bitten off,
my dog's been biting at space dust like weed-grass
because the distribution of her fluids is out of wack.
The vet gave us stuffy-nose, heavy-head,
so-you-can-rest stuff
and it works
like a face-lift:
she's already grown an inch or two.
The vet said she might pee more.
He said hearts sometimes
shrink with travel.
And when her blood pressure started rising,
he said her psycho-social needs were not being satisfied.
So I let her chase children
through the park
because her cumulative, biodynamic, interactive,
atmospheric, electromagnetic, temporospacial,
biorhythm was off.
Stress factors were,
in short,
pissing her off.
I can't tell you what that does to me.
We're both in low-gravity
water therapy now,
the dog and me. Every Sunday
she swims to me through the water balls,
her leash floating from her mouth.
She's an earthly creature
and she wants to go home.

Sacajawea, I'm Just a Fan

Writing to say I heard of you
in a song last night on Z28, Radio Utopia,
"coming to us from the red barrens of Mars County 63-06."
I heard of your Missouri River
centuries ago in the swampy backwater of our solar system.
Your grandmother told you
you would die young—
with bear grease in your hair, wild and coarsely pastoral.
Your parents were murdered by Minnetarees.
William Clark gave you your first blue coat at Fort Mandan
when he caught you kissing baby Jean Baptiste.

Sacajawea, I'm writing to tell you
there is now a new invention of you.
On Mars, we have a superstar edition hologram.
She is walking through the hem of a crowd,
doing her acoustic version
of "Kissing My Papoose."
I've made her sing that song 76 times
because that's the single I have.
She's on the sleeve of it, standing
at the canyon river Zures, blue glass
beads in her scarf, smooth mother's stone
around her holographed neck.
Your earthiness is big here—

long after Jefferson's Louisiana Purchase,
long after Meriwether Lewis is dead.
Long after Clark took you to see the mythical water whale,
you returned to your tribe and minds were changed,

explored. Sunlight from glitter and mirrors
caught in their eyes. A new world spread
like a blanket over the old.

Sacajawea, I read about you in the Martian Probe,
the space between me and statues of you—
seven hundred and seventy-seven years of history.
My curious mind is a good mind,
you would say. But you do not know
I'm just a kid working summers at Sparkie's
moon-walker lot, listening in the cab of a Crater-creeper
with the nightshift radio on—its music
breaking through all the sand devils and atmosphere.

I listen to a heavy saturation every night.
Songs extinct, fantastical,
like an old world whale, Chief Sky, the condors flying
high over the salmon running in the gorges of the Columbia.

Ute, Paiute, Hopi, Arapaho, Papago, Shoshoni, Minnetaree, Apache

Sometimes I can almost swear I see the lost people,
phantom fresh white eyes beyond the chain-link fence.
Or hear quick voices out in the dark, dodging
the crawling spotlight under the moon tires,
all debris of fame and history, fire smoke
left here on the endless frontier.
My name is Ferdinand. I'm just a fan.
Sacajawea, how empty space is
when you go off the air.

Touring Thorazine

At first, it's the light years
that set you back,
whirling in the gravity of your ego.

Then, it's the view,
even after the quarantine, after they fumigate
for bugs and mental defects,

after your mandatory appendectomy,
somewhere after the first twenty-four hours
and sixteen sunsets on your way to Mars.

We were clutching our facemasks
through the Van Allen belts of radiation,
through solar flares belching up from our sun.

They gave us tranquilizers then,
Thorazine during our free-falling back-shot around Venus.
We watched the blurry swirl of her boiling clouds

and we had to have sedatives again
while we were choking on our liquid pepper packets,
screaming out of portholes.

Tom, Dick and Harry went bad with cabin fever,
hanging in their bed sacks with the Martian Big-Eye,
while our cargo ship shuttled by with 800 pounds

of our luggage. We had missed the Lagrangian Liberation Point
and our underwear went hurling off into space.
We needed sedatives for that,

hypnotics for three more weeks of adding water
to dehydrated, thermostabilized,
packed-in-foil everything.

For funk, we needed shutting down.
And when we arrived—finally—
the ground was a tar-pit soaking up our feet.

We started counting the days all over—
twenty-five months full—
looking through the window of time,

craning our necks
down the street of stars
for the next bus back.

Failure to Launch

In every stalled car, every nose dive,
every failure to launch your heart,
lie the skids of our selves to blame.
In every affront to the foreseen,
like a rising sea of ice,
we raise our sails like ancient mariners
on our low-budget, junkyard ValueShips.
We flee wars, we seek fortune
in extra-terrestrial real estate.
We search the creator—
the Earthsink of our failed attempts:
stealing Rembrandts, legislating grave-robbing,
playing musical chairs to the death.

So Mars rises, salt and pressure
under our fingernails,
all the mountains of gold
locked in Black Hole vaults.
Mo Mars, mo problems: the crook
who misunderstands "he built this planet"
with all its universal traffic, taxes and thugs.
On every surface, the strong push off the weak.

You become the new border guard shooting ships
coming the back way around Jupiter.
You, who always intend to attend to the details;
and yet you shoot the wrong bird.
You shoot the albatross.
And since you can't believe you fell short,
you sail on—in the flutter of a dream.

You believe the bird means bad luck
instead of willful blindness and a sputtering
righteousness.

It doesn't take the long grey beard of a loon
with a glittering eye,
to see the creator gestating in a pool of carbon,
to see imagination makes the bird fly,
harmless as salvation.
He has spiritual weightlessness
and you are the sad solid swirl of atoms.

As the planets turn themselves over in the dark,
could you forget the baggage of all your ancestors,
your bell-jar mythologies?
Could you just look the bird in the eye?

…with all your grand intentions—to fuel, to fight,
to paint your face on the torso of a rocket,
to be loved in flight, to survive and to win.

Profiling Davidson, Initial Report

We know there was turbulence in apartment 00B. We know from his basement window Davidson use to watch the state of Mars today. We know how he was obsessed with fellow astronauts on leave; in particular a cliché of a smashed sailor, insufferable, swollen over café tables. Documentation shows that too many days on duty in remote outback bases, polar camps of eternal twilight, will cause anyone to feel like a planet about to fall out of its orbit.

Davidson was just a sailor in a town upset with sailors. No one says he was particularly nice. He spent six months on leave in subterranean apartment 00B, crowded with his downtown blueprints, his mail-order spear-blazers and, we now know, his homemade micro-bionic bullets.

Sources tell us he wore low hats in hardware stores. With his steady hands on the vulnerable fuses of this world, he passed the point of critical disturbance, cracked like water when the cube freezes. We know his own mind was demagnetizing him, forcing him open like a cheap toy rocket. Every night for six nights some hapless sailor left town as pieces in a box. Click, click, snap. Davidson flipped the switch with mousetrap precision, the smoke from the blow crawling up in bolero eddies. Chaos is beauty, he would say. Perfecting the timing of elevated trains coming in, this would be his finale—tipping over astronauts, the ballet of triage.

Losing at games, job dissatisfaction, music that made him tense, his motivation was incidental. His act an art, he said. Materials are what they are. In conclusion, we only know what we will never know, how to comprehend the quagmire of gravity, what art is, and where the anchor on this man's space-ship was.

Earth Watchers

This is your galaxy:
two hundred billion stars coagulating at the nipple of Juno.
Out here we get it.
We get the broadcasts from the hearts
of your third rock's greatest hacks.
We, the people of the Milky Way prairie,
we hear your pain.

You earthlings work late hours on your life's cryptographs,
your head-spinning oscillations
when something scurries up the heart.
So wracked and wrung, so eternally full,
nine quintillion miles of Andromeda couldn't cover it.
And what a poor, porous heart it is!
God squeezes it like a sponge.

God is vague and you feel dumb.
Yeah, there's a lot of God talk up here.
It's true. Out here in the galactic hood,
it's hard to dodge him. He's kind of
Fat.

Jane and Maurice

There was a strange occurrence near Indianapolis during the summer of '87. Two high school seniors, Jane and Maurice, were driving his '77 pea-green Impala through downtown Indianapolis on their way to St. Louis, Missouri. It was 7:57 pm and they were heading west on Interstate 70 with twenty-one dollars between them. Then, after a quick dinner they ran out of money, unexpectedly, at 8:09. Fortunately, Jane saw a sign halfway through the tangled interchanges of Indianapolis, a US Interstate sign that read: St. Louis—1 mile. Jane laughed and said, "Not unless we're abducted by aliens." Years later, Jane would remember this very moment, a quiet Maurice staring out into the sky. "What a great idea," he whispered. Jane and Maurice began to do what they could to get themselves abducted by extra-terrestrials: driving around in that old Impala, windows down and thumbs out in UFO-hitchhiking position. And lo, it wasn't two miles past the first Cracker Barrel, in between two cornfields, when the green Impala stalled. A shadow came down over a collapsed barn and the five-dollar compass glued to the dashboard started whirling. Years later, Jane would remember headlights going dim, the car door falling off, a green fist through the windshield and three fingers on Maurice.

For a while they sat naked reading magazines. After the ordeal of being lifted into the crowded lobby of an alien space-lab, Jane was the first to feel groggy. She couldn't focus or remember faces, only slender green hands manipulating IVs over her head. And being pressed down against a cool silver table, and blindness, and swirling air-fingers between her legs, a warm spotlight on her belly, her cervix dipping, the genteel plunging of her body surrendering ova.

Maurice remembers feeling shut out, although he received the basic abduction overhaul: hands upon hands of hairless un-identifiables with forceps measuring his long suppressed infatuations.

Over their pointy shoulders, he could see Jane spread out on a table, lip-locked with some oval-eyed Romeo. And although he would remember, on some level, feeling lucky to have escaped driving back through the long, fat farms of Illinois, he knew by the time they were returned to their proper and upright positions, he would feel lost. Rolled and diced and consequenced into the reality of unrequited love, otherwise known as bad chemistry. More than that, Maurice doesn't remember much. When he and Jane were finally deposited on the west side of the Mississippi River, impossibly at 8:11 pm, Jane was against him, sleeping in the humid early evening. Maurice would remember wondering how on earth he would be able to explain, deep in his arrested heart, the missing Impala door, a full tank of gas, and the slow blossoming of Jane's womb.

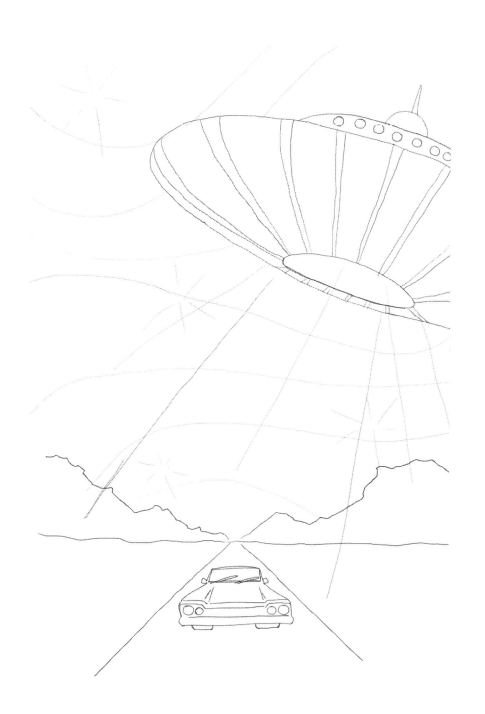

All of a Sudden

It came to me last night, the beginning, the starstuff,
the symptomatic waves of coming forth, a child.
And on her first terrestrial day, the red, waterfall
heartbeat, the game of magnets and attractions,
original motion was laid out for me.

Last night, a woman in a hospital smock laid her fingers
on the shiny belly and, mouth over face,
blew tornadoes into the water-pale toes.
Then, eyes shut and palms summoning,
my child asked me if I knew who I was.
And I said, yes, I am the speed at which
particles collide.

As I've said, the order of things
came to me last night.
First the spin of universal paraphernalia,
the quark, the anti-quark,
then physics and metaphysics and wondering
and war, the swirling fallopian Milky Way.

It was like a game of pool,
the atom of the universe that came through me.
Or a pinball, very determined,
moving movement in me, moving dolls
from dolls from dolls,
rising the curtain, a very first blink,
all of a sudden, love and creation
from mere coincidence.

Starbaby

When I was a baby, my mother called me her Martian child.
Now at sixteen, she calls me her starbaby, in the evening hours
when she thinks I'm too tired to hear and the nurses,
breaking for a smoke, snicker in the garage.
But I don't believe in highway abductions
or starship inseminations. So what can they tell me
about who my father is, from where my alienation comes.
I have such a desire to overcome this life.
But I'm not good at it. Immune to nothing,
my body has taken the path of least resistance
and my lungs argue breathing.

My mother calls me her starbaby
as if trying to remind me of my kind
of half-breed quality. But I don't believe
in sudden memory recovery, like some circuit
fully convalesced:

> *My father in curious landscapes, fire-blizzard outside*
> *hands on the strings of a low-gravity swing,*
> *scratching his sandy face, long fingers, an ovalesque*
> *mouth blowing smoke, blue ceiling, six*
> *shadows at the door, he's trying to tell me*
> *something, maybe wisdom, maybe recognition—*
> *tugging on a smooth sleeve*

What can she tell me about who my father is,
about what maneuvers he has made on my behalf:
what little alms of life, like pieces of himself
he may have left behind? Maybe his good intentions

were lost, peeled-off in space, over the barricades
he would have had to cross to reach me.

I can imagine his eyes appearing and disappearing
beyond the shadows of my bed, the missing years—
all back over my shoulder, these bones of his
holding up my body.

How did *we* come to be *me*?

In dusk—when I am alone, lost reclining
in the solemn relief of a back-porch rocker,
I'll clear my throat and the night sounds will quiet
as if listening to what I'm about to say.

Beyond the trellis, some unidentifiable presence
will speak out to my future, my possibilities.
But in the end, I'm never up to the journey,
and the nurses resume their gossiping
and I wonder who was there, caught up in the door
as I was breaking down.

When I stop breathing,
who will be there with the key?

On a Clear Day You Can See Jupiter

Some nights when the universe dips in and out of my street
like a dancer, I can see Jupiter through my window.
And I wonder where you are and how things are for you.
I wonder if writing poems for me is like painting for you,
like performing resuscitations on a dream.

And although we are together—fundamentally here
on the same hemisphere, you don't have to answer me.
You don't have to reply to this untethered planet heart.
It's too late for us and I surrender to the war
of my fates—where poems burn into pieces of litter.

Take your watercolors and color my window with Jupiter.
Crack open the glass with your knives and turpentine.
Paint these words of mine, life-full of hue and value
and watch my heart healing like a bone
inside the tornado of a thousand bristles.

You don't have to answer me. You just need to know—
on a clear day you can see Jupiter—in my eyes
and on your fingertips, where the universe
dips in and out of your street like a dancer,
in my words and through this window where I'm on the horizon.

Maribel, Crossing

If I am a comet and you are a planet we call Johnny,
and I get trapped in your orbit
and ripped apart by your gravity,
(disassembled and depressed
wherever in the universe I am),
If I am constantly circling
and the Bureau of Astronomical Vehicles
calls to say my course is erratic
as I am always evading civil collisions with smooth-faced meteors—
(unbearably immaculate gentlemen),
If I am rushing at you from the furthest arms of this solar system
and I cry, *Johnny, come lightly*—

you should know I couldn't slow down.
I tried. I was young. I was proud. I was sure.

Helga Post-Orbit

I have my doubts about being dead.
My dog was sucked out into space this morning.
Isn't that what happens,
you get sucked right out of space?

Flung from the nine coy planets,
it must be like solitaire and hunger out there,
awake and not knowing where you are.
Is she waiting for me to open yet another door?

Or has she trotted off with the other dead?
Has she ceased to be anything
but synapses in the back lot of my memory?
It's not the voided space she left behind

that bothers me. That kind of affection
can always be shifted
like square-pieces in a pointless game.
No, it's not the sentimental, leftover space

that matters as much as the idea of Helga, open-eyed
and drifting, somewhere out of the room.
Or this mortal part of me—lost in a raw,
everlasting free-fall of disconnected, disordered love,

knowing I'll uncover Martians, (Martians!),
the impossible mysteries of Mars,
before I'll ever know
where Helga has gone.

Children of Algebra

Novalis said "the life of God is in mathematics"
and I am the mathematician, a sensitive,
predisposed to trust initial conditions
as they flip down the click-clack cosmos
like a neck snapping from a mistaken thrust
on parallel beams, or a hurricane blossoming
from a sigh in San Juan. I am your teacher
but can I teach you to climb trees again?

You must have faith in irrational numbers,
the sum total of what you have become,
find formulas in the chalky sky,
the spiral graphs of seagulls, cheeky geometry.

Children of algebra, you must work
through my binary lies with your questions
and your hypotheses of discontent—
all your lives adding up to fractals.
Count on your fingers—if it calms you.

I apologize. I am the absent teacher
gone for most of the hour;
I only really wanted to coach
the games of bees and bears
in this collapsing, expanding lung of a place.

If you want it, this is your composition assignment
in one hundred words: to mourn
your inflatable fate, punctured rudely
by a bitter meteorite
who tore through your paper proofs
just because he could.

Um Is a Wonderful Word

Two monks are sitting in a bar,
one with his robe hanging ajar.
      No,
      wait a minute,
there was a nun in there
      too
at a booth with a plastic bowl
      of peanuts
      and a screwdriver
      and a foreshadow
under the table.

Under the orange and vodka ceiling
of her glass
      a crushed
      and sterile cherry
      prediction.

Slipping through the loopholes,
though the laws of conceivability,
      outside—in the suspended night
three dolphins named
Tetrahedron, Anomaly and Brake
float across the window
as chrome corkscrewing through the midnight.
A bartender drops his books
      of course
      of course

the ones he's been stealing out
to his correspondence course;
and the waitress who swivels on a stool
rclinquishes her arms
        of Presage, the peach cat.

At the end of the bar,
a door full of bottlenose dolphin
swimming hot tail,
impossible stealth
tricks of physics,
surfing the bow waves
        of space
            ships.

Mother Nun turns to her bartender and says
        "I hope you understand,
you will be rewarded for your hard work."
And somewhere Tetrahedron, Anomaly and Brake
make their last dive through the stars.
Their lungs collapse
            promise.

Back in the bar,
standing on top of a table
near the bathroom,
an amateur philosopher, a poet
is trying to explain himself.
        Um... is a wonderful word.

Monogamous Carbon: a Classified Ad

You can come by my pinball planet, carbon-boy,
and start something
like love, if you can imagine
my Mother Earth's lazy evolution.
I've been ten billion years of waiting
alone on Saturday nights,
sweating for those essentials you've been collecting:
three buckets full of indestructible energy
and the principles of self-assembly.
Meanwhile, you've been years carousing,
a tide-pool bubble trolling undisturbed
in the foreground of Earth's day-to-day burning violence.
Come over if you can caress me like a plasma
around sharp rocks; which is, incidentally,
what my father will not like about you,
but what will cook and weld through the years,
via volcanoes and asteroids and, frankly,
dinosaurs melting from here to then,
progressing, failing, melding
into your human silhouette at the door
by the lava lamp
with a child in your arms
and all the flowers
just details
scattered along the way.

Things My Inner-Astronaut Is Afraid Of

- Inter-stellar confrontations
- Rorschachs in the hearts of swimming pools
- Barroom brawling and spaceship psychology
- Peacocks with asthma and going first
- Being left behind as the planet makes a right turn and swishes off into the black, sequined yonder
- Losing ground
- Blasphemous religions of intolerance
- Manifest Destiny
- Animals trapped on the wrong side of the freeway
- The poet lurking in all of us

The End, Part III

So this is really the end, is it? Time to waltz with your ghosts, watch your life coming apart, particle by particle, your DNA falling off its ladder.

What a blow to regret and revision.

A shower of meteors pocketing your planet or a comet slamming in and shaving off the surface or two thousand asteroids escaping the gravity belt, apocalypse, a plague in space, random entropy, the sun sputtering out, God's own sovereign TNT.

It doesn't matter anymore.

Time bended. Time ended just like we knew it would: you can't see the fourth dimension; you can't see back in time. You can only move one way. Otherwise nothing would hold together, no cause and effect, no crater-creepers, no leftover dinners, no mailbox Polaroids or jungle-gyms.

It's only a matter of time before the birds and the brouhaha, particles of arithmetic, history and biology, the half-breed soul of reality shows you—

the indescribable shape of your stopping.

Y

If you're wondering why I'm still here
how I am
how I am
still here
a soul preserved in amber
If you're out here looking for a voice
spooling up a choice
you made back then
when you were
who you were
If you still don't know
who the hell I think I am
to be packing up this existence—
for the time being
just being time
If you're still wondering
how it is I
how it is I
survived
I survived
I let go of my x y z world
I let go of my stake-through-the-heart
Why

Why Photographers Commit Suicide

When I think of all the things I've seen on Mars,
I think about the photographs
Maury Etc took,
those grainy prints still hanging
on the gallery walls of Astasia,
our first human colony.
His silver signature, cross of the lonely t,
is still framed there in rare earth wood,
set behind the slowly moving glass.
"You have no control," he said,
"over what you see.
A photographer can only function as long as there is light."

When I speak of things I've seen on Mars,
I always mention the *Badlands Series*—
not a single person in them, just his land rover
behind the volcanic vents;
not a single Martian in them, no shadow,
just the ancient river channels, the tiny tornadoes
across the rusty-brown flatlands, the mesas, the crystal fracture
of Valles Marineris, the canyon five miles deeper,
two thousand miles along space and time.

I mention the Dome of Tharsis—
"There we shuttled over Olympus Mons," he said,
"tallest volcano in the solar system.
It's strange," he said,
"there's just some missing thing."
He touched the water in the holding bath,
pointed to the picture, the volcano's caldera

forty miles wide.
"Some ghostly thing called flow."

 "We found no traceable remains
on the ultra-singed surface," he said,
"but I could see faces," he said,
"looking out, faces
looking in through the slow-shutter blur,
orange haze, rolling dust.

"I saw them standing in the water that once was,
my human eye filtering for blue under an over-
exposed sky. I saw the subterranean
glowing eyes of the last breathing Martian child
closing up the house in an incident of light.

"I saw creeping fire clouds, the millennium flash,
heaving crumbled heaps of them smothering
steep down Sacramento Street,
the burning, the dodging, the trapped light,
fifteen minutes of rainbow
losing atmosphere.

"But on metallic silver, in the sensitive cell,
in the fixer, in the tones, on the plates and in the pictures,
not a single person in them,
just sand dunes running
at half the speed of sound,
extinction in the foothills,
strange marks on the surface.

"I tell you, boxes don't take pictures.
There's the human filter factor,
the leaf-shutter eyelid of the retina,
spotting, mounting out of context,"
he said, "always out of context.
Listen, you just can't visualize the fourth dimension.
Even with infrared, you can't see back in time."

The last time I saw Maury
he was setting up a rusty-red flag
at the Mangala Valles horizon.
"There is no absolute point of rest in the universe," he said.
The turbulent rolling tarp looked wet on the red planet.

And so without a headpiece, without even space shoes,
he wandered off into the morning,
bouncing along the sandy conscience
into the low thin weather of depressurization and death,
leaving very small footprints in the dust.

Acknowledgements

My gratitude to the following journals and presses who published earlier versions of these poems: "Why Photographers Commit Suicide" in *Natural Bridge* (1999) and by Hilltop Press (2004), "All of a Sudden" by *Tintern Abbey* (1998 online), "Imagine Mars" and "Sacajawea, I'm Just a Fan" by *The South Carolina Review* (2007), "Traveler Beware" by *Phoebe: Journal of Gender and Cultural Critiques* (2008), and the Helga trilogy and astronaut list poems in *Eye Dialect* (1999 online).

I am also grateful to the brilliant teachers of poetry who helped shape these poems: my first mentor, Howard Schwartz, and Steven Schreiner (and the amazing foundation I received at the University of Missouri-St. Louis), Thomas Lux, Jean Valentine and David Rivard (who all helped me through an early version of this book at Sarah Lawrence College), and my compadres who provided astute and crucial commentary: Julie Wiskirchen, Christopher Brisson, Sherry Fairchok, Murph Henderson, Joann Smith, Denise Vacca and Marilyn Koren. Thank you to Ann and Michael Cefola for encouragement and lessons on astronomy. Thank you to EM635 at Scribendi. I also want to thank my pottery teachers who guided me towards a way of thinking that exists beyond words: Bruce Tomkinson and Linda Mechanic (in Los Angeles) and Adele Devalcourt (in Santa Fe).

My thanks to the talented artists who helped me put together this project: Stephanie Howard, Marilyn Ghigliotti, Jeff Ytell and Emi Villavicencio.

Final appreciation to my parents, Dave and Estelene, for their smarty-pants genes and for giving me all those childhood opportunities to read and ponder stuff; and all my love to John McCray for our soulful conversations and for his interminable love and patience.

About Mary McCray

Mary McCray is the author of *St. Lou Haiku* from Timberline Press and co-creator of the web-zine *Ape Culture*. She also blogs about pop culture as *Cher Scholar*. Mary has poems and essays published in journals such as *Phoebe: Journal of Gender and Cultural Critiques*, *The South Carolina Review*, *Literal Latte*, *Mudfish*, *Book/Mark*, and *Hermenaut (The Journal of Heady Philosophy)*. She lives in New Mexico with her husband, archaeologist John McCray, and their two fur-kids. To connect with Mary visit her online at www.marymccray.com.

Twitter: twitter.com/mary_mccray
Cher Scholar: www.cherscholar.com
Cher Scholar Blog: cherscholar.typepad.com

Visit the poetry blog *Big Bang Poetry*, reinventing the life of the poet in the modern world:

www.bigbangpoetry.com
Facebook: Big Bang Poetry

About Emi Villavicencio

Emi Villavicencio is a graphic designer who loves to work with color, texture and freehand illustration. She grew up with a strong love of art and received her bachelor's degree in graphic design from the College of Design, Architecture, Art, and Planning (DAAP) at the University of Cincinnati. She currently resides in Cincinnati, Ohio with her family and friends. To learn more about Emi, visit http://be.net/emiv.

Hometown Haiku

St. Lou Haiku

"If you find that the universe manifests itself more in toasted ravioli than in the plight of the mongoose, you're not alone."
—Kristie McClanahan, *The Riverfront Times*

"Ladd and Wiskirchen take on St. Louis attractions, personalities, work, food and summers with an interesting mix of sentimentality and biting wit. From the "zombies" at Union Station ogling the fudge-making process to the haughty suburbanites who forsake a great meal on The Hill for dinner at the nearest Olive Garden. St. Lou Haiku pulls no punches." —Matt Berkley, *St. Louis Magazine*

This first edition of *St. Lou Haiku* was handcrafted in 12 pt. Garamond Old Style type which was handset. Printing was done on a 6X10 C&P hand press. The paper is Wausau Royal Linen with Royal Silk scarlet endpaper. The zinc-cuts from Clarence Wolfshohl's four original St. Louis illustrations were made by the Augustine Company of Marshalltown, Iowa.

http://www.stlouhaiku.com

The Prairie Traveler

More poems by Mary McCray

A handbook for pioneers, drivers, explorers, homesteaders, settlers, drifters & exiles. Includes recommended routes, aid, shelter & wagon maintenance.

Desire, to Begin

(Formal introduction to this guide)

Maybe you hold this book of pages, this document, your hope in your
    hands
because you are fighting at the bit in a fire-cold, restless suffering
and you look for the wisdoms of my odyssey in the desert;
you need a guide to help you survive a Go West, what I went through
those years ago when I was young
and wanting to be a fine and famous journalist,
and wanting to be understood.
You may learn, perhaps more quickly than I have,
and far more comfortably, by your fireplace smoldering with risk,
sitting on your life's fine-strung chair with your cabinets full of what
    you have
and do not have,

the taste of sweet, imperfect melancholy.

Maybe you seek the physical trials, the thrill of the trail,
a mountain's stark understanding of sky sovereignty,
the coarse dust of the brush, the sweet smell of the spring,
dirt-boot experience, sleeping out in the open.

You seek a Master who has traveled into the famed prairie lands,
someone who can take you through the spirit road,
gently and roughly,
until you too become bronzed, creviced and calm,
high-desert-holy, canyon-marked.

Hold out your tin for the divine elixir.

Take the breath of atmosphere from that cup
before you head down the trail of my tale,
before you learn how to rope a cow, how to cook beans
and the cake of cornbread over a fire, how to ford a river
how to cross over.

How to ask yourself—is this my call to ride?

But be forewarned: this story is mine and only mine
and never really was mine.
This is the way
of the journey.
I am only one lost and found journalist…unraveling and unraveling
his basket of truths.

Swimming the Herd

(How to think like a cow)

We spend our young life
mocking cows
and how they move like chaos,
plodding, milling, bawling stragglers.

The smell of shit, a dumb brute,
the dusty lunkheaded mob,
vacant, vapid heifers,
stubborn, mindless,
ungovernable mass.

Their quiet unconcern as we count, recount
and tally our anxieties.

His thinker wouldn't make a tea cup
for a hummingbird.

Nothing dumber than a cow,
except for the man that herds them;
it is the futility of a gnat
biting an iron bull.

At dusk they are a classic outline of balance,
centered into the earth.
Vaquero Joey sings Verdi, *La Traviata,*
like a lullaby to the herd.
They gather bemused,
fleas buzzing above their humps.
Calm brahmans, the long tongues
of longhorns staring out to the horizon.

They smell tomorrow's rain;
They side-eye your circling approach;
They stare at your audacity in disbelief,
chewing their cuds, re-digesting the facts
of each blade of grass
four times.
They know you lie.
They know you lie to yourself.

Mindless? Then they are what you seek,
your cattle-koan to solve.
Listen to their low bellowing moo,
the Om of their meditation.
Enter the herd bovine.
Lay down on the plains
and dream of water.

Tom is Dead
by Mary McCray

In this collection of poems, Mary McCray explores pop-culture life, Hollywood and celebrity as the modern-day fairy tale. She takes on Katharine Hepburn, Captain Hook, King Kong and other divas. "Orgasmic Orange" was originally published in *Switched-on Gutenberg* (2001 online), and "Scheherazade on the Landing" was originally published in *Mudfish* (2009).

Orgasmic Orange

Maybe you were once a cowboy.
Your last life, you were Joan of Arc,
Valentino or Edgar Allan Poe.

I was thick-skinned and full of water.
I was beautifully balanced and well round.
I was wimpled and mandarin and blooming buoyant.

I was bold and I was bursting.
I was hanging in the valley of Apollo,
Puckering and zealous for tongues.

Scheherazade on the Landing

The Sultan King is standing crooked at the bottom of the staircase
clutching a bat. Just like Jack Torrance in *The Shining*,
he says, "Scheherazade, you've had your whole fucking life
to think things over. What good's a few minutes more
gonna do you now?"

Scheherazade hovers over her cockamamie scheme,
her marriage, like it was the custodianship
of a haunted hotel. She can handle the emotional distance.
She never cared if all work and no play made Sultan a dull boy.
It's just his aggressive condescension she finds hard to take,
the sarcastic tone he assumes when he backs her into a corner,
the way he says "light of my life" as if she's not really so light after all,
the way he steals her time to think as he backs her up the stairs
toward the monstrous, hanging carpets,
backs her up over the precipice of the landing
while she's flailing her arms in an effort to protect
a glass heart against a man with a bat.

This is the beginning of the end, right here—
her Sultan turning the corner with madness,
primal, animal madness—his love.
Why does it play this way over and over again,
this trek up the stairs to destroy what he loves?

How is it that he has become the mortal enemy
of his own glass heart?

Big Bang Poetry

Reinventing the Life of a Poet in the Modern World

Welcome to Big Bang Poetry

Poetry matters. Poets matter. There is a place for poets in our culture. It's time to start thinking outside the box, even if you hate that phrase. This weekly blog covers:

- Books to read
- Commentary on craft
- Interviews with writers
- Action items for poets
- Poetry news
- Support for writers

This blog is a challenge to think outside the lament. Think outside your own rationalizations about who you are and what you do as a poet.

Cher Scholar

http://www.cherscholar.com

Cher Scholar has pontificated academically about Cher for radio stations such as NPR and for publications such as the *Las Vegas Sun,* the *Philadelphia Inquirer* and the *Winnipeg Free Press.* Visit the website for interactive games, Dear Cher Scholar, reviews and interviews with Lenny Roberts and Rona Barrett among others.

The Weekly Blog

http://cherscholar.typepad.com

Launched in 2006, *I Found Some Blog...by Cher Scholar* dissects both fan culture and Cher's multimedia career in film, music, television, live performance and as product pitchwoman. Topics include Cher in Arts & Literature, Outfit Watch and That's the Obsession Talking.

The Zines

http://www.cherscholar.com/zine.htm

"A tribute zine in its finest form. Touching and humorous writing from true fans of a cultural icon." —Kristy Mangel, *Zine Guide*

"A must for Cher fans." —Anu Schnuck, *Zine World*

"This zine did the near-impossible for me, it made me care about Cher." —Dann Lennard, *Betty Paginated*